All in the Family

A SCIENCE IS WHAT AND WHY BOOK

All in the Family

ANIMAL SPECIES AROUND THE WORLD

BY GILDA BERGER
WITH DRAWINGS BY TOM HUFFMAN

COWARD McCANN & GEOGHEGAN · NEW YORK

General Editor: Margaret Farrington Bartlett

 Published simultaneously
in Canada by Academic Press Canada Limited, Toronto.

Library of Congress Cataloging in Publication Data
Berger, Gilda.
All in the family.

(A Science is what and why book)
Includes index.
SUMMARY: Describes the multiplication and
distribution throughout the world of various species
of cats, camels, bears, and humans from their
ancestors millions of years ago.
1. Zoology—Variation—Juvenile literature.
2. Zoogeography—Juvenile literature. 3. Species
—Juvenile literature. [1. Zoology—Variation.
2. Animal distribution] I. Huffman, Tom.
II. Title.
QH408.B47 591 80-17803
ISBN 0-698-30730-5

First printing
Printed in the United States of America

To Herb, in fond memory
—GB

To Marcy, Gary, Willie, Debbie and Linda
—TH

Do you know that there are about fifty different kinds of house cats? Among them are calicos, Siamese, Abyssinians, Russian Blues, Burmese, and many types of tabby cats. Some are short-haired; some are long-haired. They are brown, black, gray, white, solid, striped, or a mixture of colors. Some are fat; some are thin.

In spite of their differences, all house cats are closely related. People say that they belong to the same *species*. A species is a group of animals that have similar body structure, eat similar foods, and have similar instincts. Above all, animals of the same species can mate and have offspring. A male and female house cat can mate at certain times and produce a litter of kittens.

There are other kinds of cats that are not house cats. Lions, tigers, jaguars, leopards, ocelots, and bobcats are some of the other kinds, and each one of these is a species too. The species of house cats is related to other species of cats. All together there are about thirty-five species of cats.

The various species of cats make up the cat *family*. The members of a family are similar in some ways. But there is one big difference. Members of different species in the family do not mate and produce offspring in nature. A leopard does not mate with a tiger; a jaguar does not mate with a bobcat.

Species in the cat family live far and wide. Lions live mostly in Africa, tigers in India, jaguars in South America, and bobcats in North America. They are spread out over most of the land on Earth, but this was not always so.

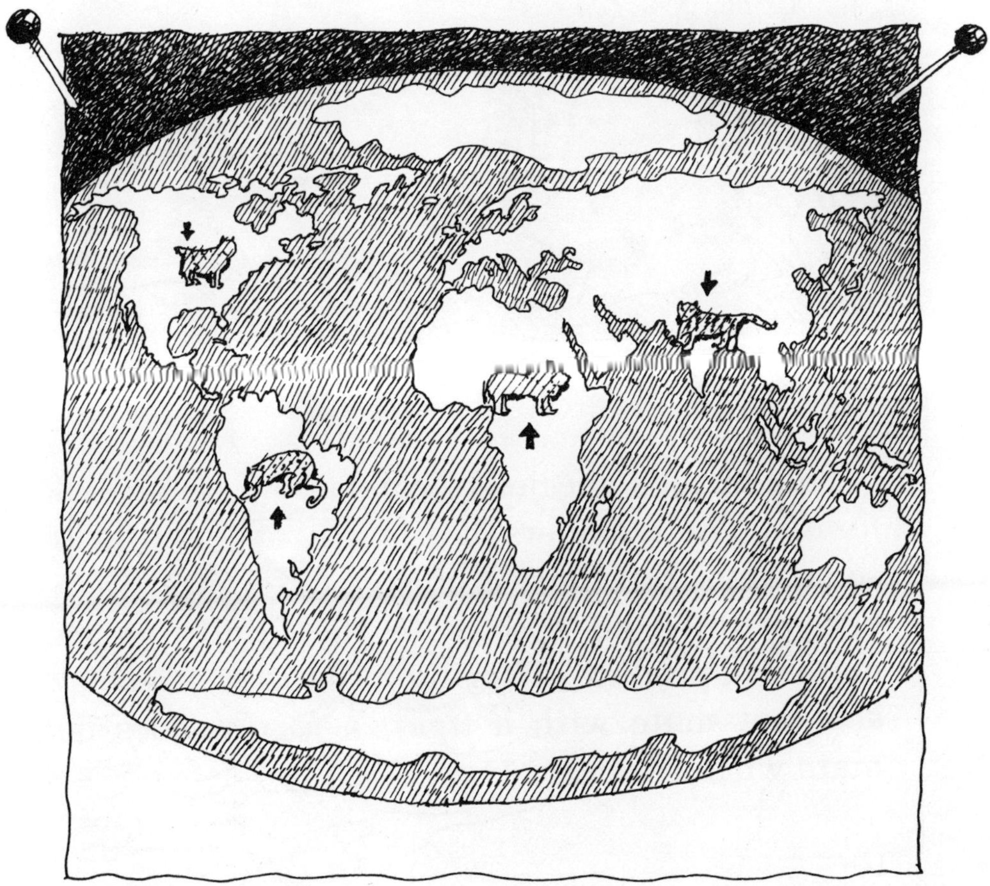

Millions of years ago there was only one species of cat. This was the common ancestor of all members of the cat family. Most likely it was much smaller than most of today's cats. However, it could pounce and capture other animals, and it ate meat.

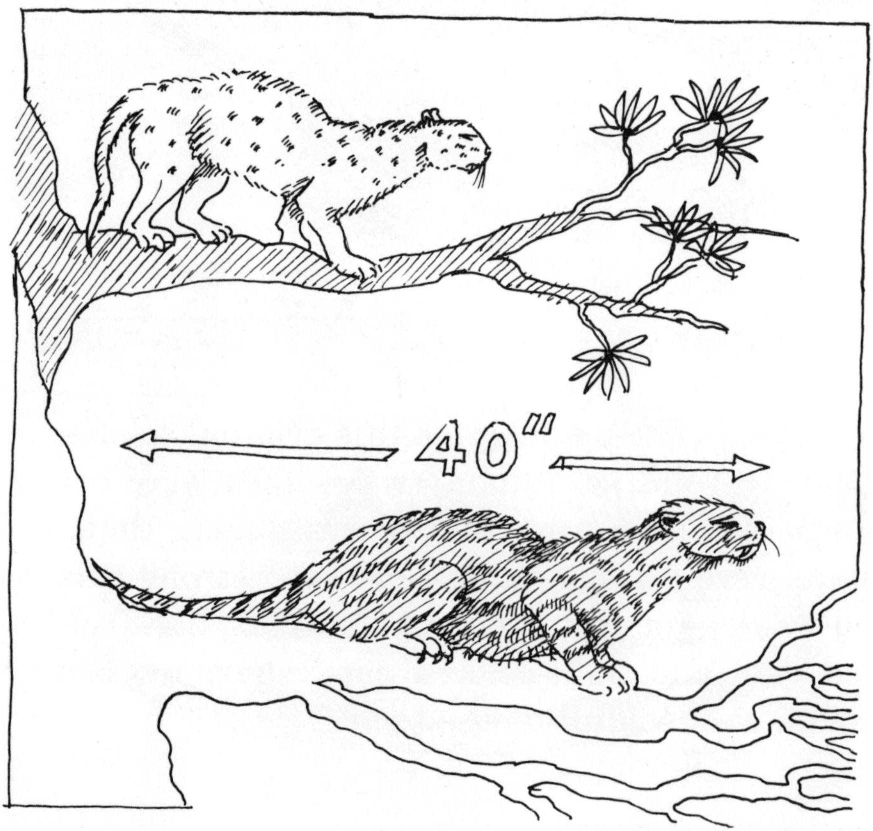

Although the members of this original species of cats were alike in many ways, they were not identical. There were differences among them. There were tall cats and short cats, strong cats and weak cats, fast cats and slow cats, dark-colored cats and light-colored cats, sharp-toothed cats as well as dull-toothed ones.

In time this species of cats spread out and settled in faraway places. Each place had its own set of living conditions.

The animals now had to fit into forests or jungles, plains or deserts, hot lands or cold. In some places the stronger, bigger, cats were better able to get food. They could better escape their enemies. These animals survived. They had offspring. They passed their special abilities for survival on to them. In other places the more slender, smaller, faster cats were better adapted.

Those which did not fit in as well were not able to get as much to eat. Many were killed by enemies. Therefore, they produced fewer offspring. After many generations, fewer individuals in the population had the less valuable features for the environment.

All the while the earth itself was changing. Earthquakes and volcanoes were forming new mountains and valleys. Rivers changed their courses; new streams and lakes appeared. There were big changes in climate and rainfall. New forests and grasslands sprang up. Even the continents slowly drifted apart.

Some groups of animals were cut off from each other by changes in the land. Mountains, forests, and rivers separated them from others of their kind. Animals in each group changed until they were so different that they did not or could not mate with animals in another group. They could only mate with others of the same group. Although they were still in the same family, they had become new species.

One form of cat became fit for living on the open plains of Africa. Its dark tan color blended in with the color of the soil and rocks. It was a good hunter. It could spring up to twenty feet to attack. With a blow of its foreleg it could kill its large prey. The lion came from this line.

Another form became fit for living in the jungles of Asia. It hunted for food on the ground. It used its sharp teeth, powerful jaws, and slashing claws to capture and eat small animals. It was able to swim, and so it could move to places that had better hunting. From this line came the tigers.

Some dark-colored cats became fit for living in the rain forests of South America. They were good at climbing trees. They could pounce on their victims to capture them. These animals were large and strong. They became the modern jaguar.

One group of small cats ended up in northern Africa. The Egyptians, in particular, fed these animals and kept them in their tents and in their homes. After a long while, they tamed them. They used the cats to catch the rats and mice that were eating the grain that was stored in warehouses. From this line came the house cat, which so many people keep as pets.

Today's bear family includes eight different species. Among these are the five-hundred-pound black bears that live in the forests of America.

Then there are the even larger brown bears that live in the woods of Europe, Asia, and North America, and also the very small sun bears of Burma, which are about the size of dogs. They weigh less than one hundred pounds.

All species of bears came from the first bear-like ancestor, which appeared on the combined continent of Europe and Asia about 13 million years ago. By the time of the great Ice Age, some one million years ago, the species of brown bears had a wide range in both of these continents and in North America.

As glaciers advanced over the land during the Ice Age, the climate changed. Many of the brown bears could not find enough food to eat. They no longer fit into their changing environment. Some wandered away from these harsh conditions. One group came to live along the northern sea coasts where the climate was a little milder.

This group of bears became isolated. Since there was not enough food on land, they turned to the seas. The better swimmers found more to eat than the others. Those that had light-colored fur were harder to see on the ice, and escaped their enemies. Those that could stand the icy waters did not freeze or suffer from the low temperatures.

Eventually there came to be a new species of bears, the polar bears. Polar bears live near the sea. They have relatively smaller, more pointed heads and longer, thinner necks than most other bears. This shape helps them to move quickly through the water and to catch fish while swimming. They also have better eyesight and sharper teeth than the bears that feed on plants that grow on the land. Although they are part of the bear family, polar bears are a separate species.

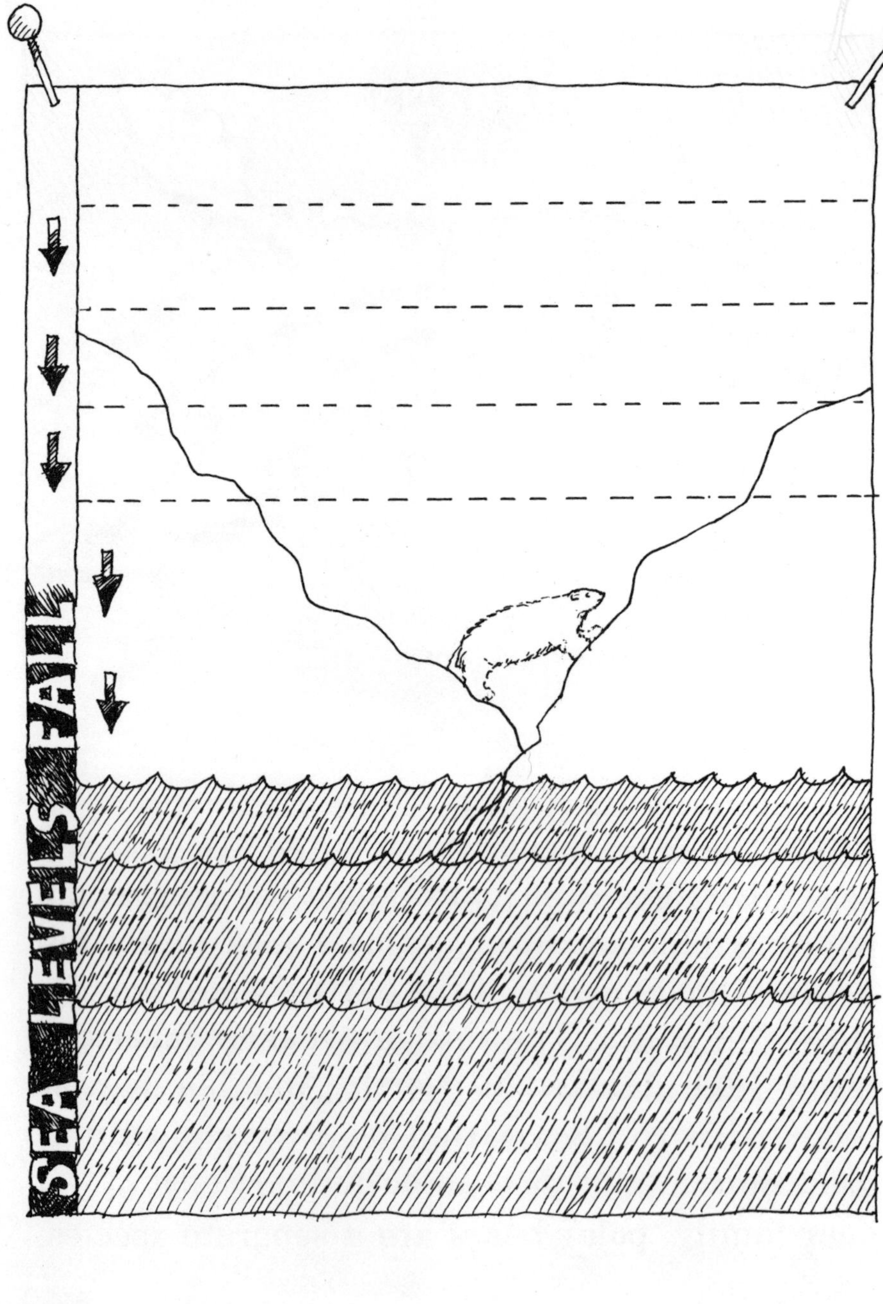
SEA LEVELS FALL

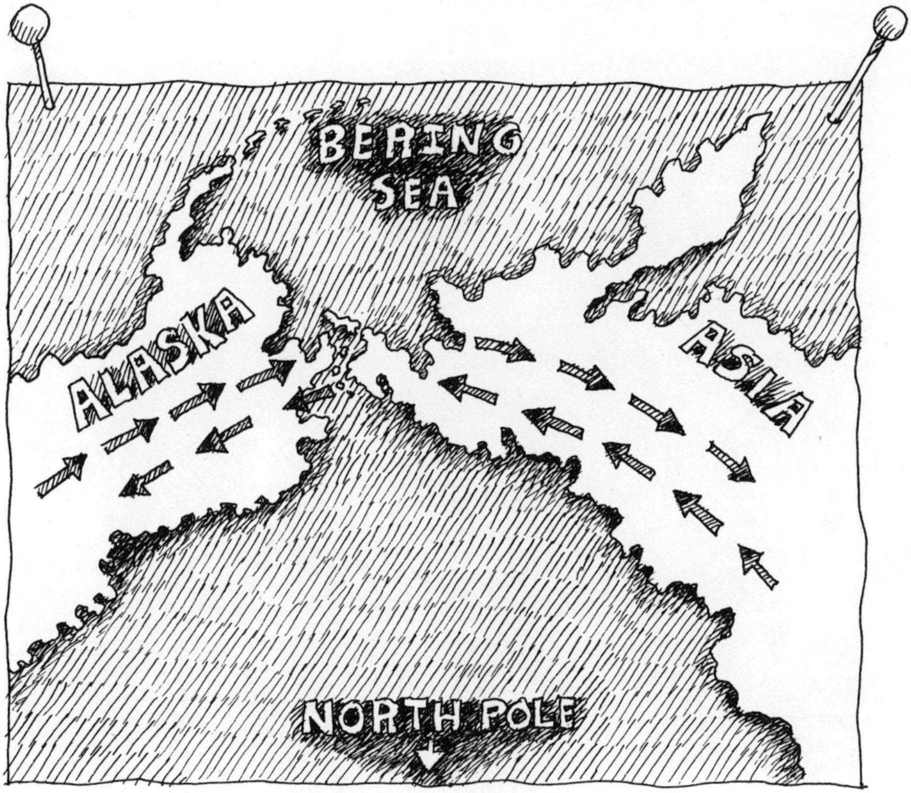

Land changes during the Ice Age also removed some barriers that had divided the ranges and stopped the spread of animals. Sea levels fell, since large amounts of the earth's water was frozen in glaciers. The bridge between Siberia in Asia and Alaska in North America became wider. It was then easier for groups of animals to cross these bridges in both directions.

llamas

Camels were one group of animals on the move. They started out on the plains of North America. Some migrated over a land bridge to South America. Among them were camels that had long ears, small feet, a short tail, and no hump. The present llama of the Andes mountains was derived from this sort of camel. The little vicuña and the larger guanaco of South America are other close relatives.

Some camels migrated to Asia. Among them were camels with humps of fat on their backs. Some became fit for living in desert areas. They could go for long times without food or water. They are today's one-humped beasts of burden, the Arabian camels.

Other camels, with two humps, were better fit for life in the colder, rocky hills of central Asia. They are the modern two-humped Bactrian camels.

Oddly enough, there are now no camels native to North America.

Why they became extinct, though, is a mystery.

The llamas, vicuñas, and guanacos of South America, and the Arabian and Bactrian camels of Asia differ in many ways. Yet, they are members of the same family. Their bones and body structure are alike. They all have heavy coats of wool. They are strong and can carry heavy loads. They can run quite fast. Each one belongs to its own species.

There is only one species of humans today. Fossils of a few million years ago, though, show that once there were several man-like species. They came from a common ancestor. There was probably fighting and struggling among them. Only one line survived. That line led to our present human species.

The first of our species may have arisen in northern Africa about 500,000 years ago. As they became more numerous, some groups moved out. Some went to live in caves in South Africa.

Others journeyed northward. Some set out in canoes and settled island after island until they reached Australia. There were people who went east to Asia. From these settlers came groups who crossed to North America and eventually to South America.

A long history of living apart led to some differences among people that were helpful for survival. Those with lighter skin who settled in northern Europe fit in better there. A light skin helps the body to produce vitamin D in places where there is not much sunlight. Dark skin is better in the tropical areas of the world. It protects against the sun's powerful rays. A skin shade between light and dark is best in Asia, where the climate is between that of northern Europe and tropical Africa.

But living apart did not isolate groups of humans. The barriers on earth did not separate groups of people as they did other groups of animals. Humans were able to cross almost every barrier and fit into almost every environment on earth.

Today human beings vary in about as many ways as any species on earth—even house cats. We vary in skin color, body size, body shape, hair texture, eye color, and in many other ways. Yet, in spite of these differences we belong to one species. We all have similar skeletons and organs. Our bodies work the same way. We have similar instincts. And we can mate and produce offspring.

The species of cats, bears, and camels, and the human species are only a few of the million or so different species of animals that exist today. All together, they make up in great part the richness and fullness of life on earth.

INDEX